GOLD FEVER

The Art of Panning and Sluicing

By

Lois M. De Lorenzo

Copyright © 1970
Gem Guides Book Co.

Revised Printing 1978
Fifth Printing 1983
Sixth Printing 1985
Seventh Printing 1987
Eighth Printing 1989
Ninth Printing 1991
Tenth Printing 1993
Eleventh Printing 1995
Twelfth Printing 1997
Thirteenth Printing 2002
Fourteenth Printing 2005

Gem Guides Book Co.
315 Cloverleaf Drive, Suite F
Baldwin Park, CA 91706

Dedicated to my husband, Frank, and to my three sons, Rich, Mike and Chris who share our enthusiasm for hunting gold.

and

To all new-comers to this hobby: May your pursuit of gold add new dimensions to your outdoor hobbies and bring you good health and much happiness!

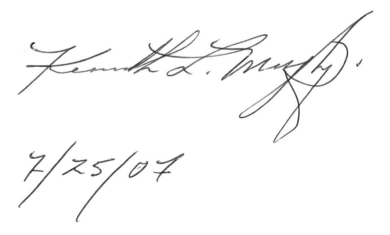

7/25/07

Library of Congress Catalog No.
79-90416

ISBN 0-935182-00-4

FOREWORD

GOLD FEVER is a disease that has excited man's imagination since the dawn of time.

GOLD FEVER is a state of mind, a happy mixture of matter and an idea.

GOLD FEVER lasts indefinitely, and its victims are never completely cured.

GOLD FEVER is contracted the moment you find your first fleck, and it will awaken in you something that cannot be explained.

GOLD FEVER is a sensation that is pure and honest, because you know you earned your reward by hard work.

GOLD FEVER is contagious. You can catch it from other GOLD FEVER victims.

GOLD FEVER symptoms are not fatal, but there is little hope for a cure.

WARNING!

Reading this book could infect you with GOLD FEVER and you may never be the same again!

MUSEUM

COURT HOUSE
1866

MY FAVORITE TOWN

LOIS DG.

CONTENTS

HOW IT STARTED

Our escapades in hunting the elusive GOLD started in the Mother Lode Country of California, but it could happen to you no matter where you happen to be.

We were "just passing through" on our way home from Oregon, along Highway 49, when we were caught up in the fascinating atmospheres of the old-time rip-roaring Ghost Towns that had settled down to quiet and more peaceful ways.

I became intrigued with the dilapidated structures that had turned silver-gray by the ravages of wind and rain; but enjoyed their counterparts more. . .in towns where the buildings had been lovingly painted and cared for, from the tip of their gingerbread ridge poles to the lacy decorations along their rambling porches.

What stories they could tell!

Walking into hotels which had been stage depots or landmarks in bygone years (refurbished with antiques, gas lamps and lanterns and Victorian fixtures) gave me an eerie feeling of having stepped over some "twilight" boundry and back into the past.

We slept in hotels, where rusty, hydraulic mining equipment adorned their landscaping; ate meals on a patio built out over the river, while admiring the old homes and buildings that peeked at us between the tall pines that marched up the mountainsides; had hamburgers and fries served to us on gold pans; walked over rough wooden walks, and stepped on hollowed out boards that had counted many footsteps.

Exhibits in museums brought to mind stories of the Forty-niners: old wagons, primitive cooking utensils, hand-made tools, "six-guns," cemetery markers, and pictures of tent-towns that had sprung up overnight (and disappeared just as fast when the shout of "GOLD!" stirred hearts to move on), but . . . displays that held GOLD . . . held MY attention!

My imagination exploded as I beheld samples of GOLD, GOLD in quartz, pyrite with GOLD, "wire" GOLD, and GOLD NUGGETS!

On the last day of our vacation, I saw displays of shiny, new gold pans and shovels in store windows. When a store-keeper told me gold could still be found in the river, I could resist the impulse no longer. I bought a GOLD PAN!

Of course, I was eager to try my luck. Frank wanted to fish; so we drove back along the river-road, found a place to pull off, and parked.

My husband decided he'd have "ABSOLUTELY no part of this foolishness" and took off with his pole and headed for deep water. I started to dig in the sand.

Time after time I filled my pan and washed it out and tried to recall the brief instructions given to me from the store-keeper; but not knowing for sure how to hold the pan or how to "slush" the water around, I began to get discouraged. Maybe Frank WAS right. Maybe the river HAD been cleaned out.

I climbed the bank and tried different spots and had almost given up when I came across a fallen tree. Sand and stones still clung to its roots so I scraped some into my pan then rushed to my "cove" to wash away "the top stuff." When I was about to dump the last of the sand into the river, I saw a glint of "color" trailing behind in the black sand.

I yelled in delight and Frank came running and splashing to see what was the matter.

"How'd 'ya know THAT'S gold?" he asked.

"The fellow in the store said it would be the last thing left with the black sand. . . and it is! And . . . it looks like the gold we saw. YUP, it's the real thing!"

He mumbled "Beginner's luck" and went back to fish.

As I fingered those small specks, I felt the same tingle of suspense and excitement that had urged the early miners to go on hunting. I'd found GOLD!

With more enthusiasm, I went back to work.

When I looked up to see how Frank was doing, he was nonchalantly kicking over some rocks in the river (while retrieving his line) and I knew my few specks had captured a corner of his mind, too.

Soon he came back and said he was "taking a break" but after watching me a while, asked, "Let me use the pan and try a spot?"

I could tell that fishing had lost its hold on him and he was getting "hooked" on GOLD.

Well, we dug sand and gravel all afternoon until sunset, and it was time to return to our motel, to clean up and be ready to move on in the morning . . . but we talked all evening and all the way home of *OUR* gold and worked out a plan to return "next year."

* * * * *

During the months that followed, we collected old spoons, knives, screw drivers, whisk brooms — anything and everything we could find that would be usable to clean out crevices, and stored them in a box called "FOR GOLD PANNING."

Whenever we could get away for a weekend, we ventured out into the desert with extra water and a wash tub (to take the place of the river) and tried our luck. Using maps of gold-bearing areas, we looked for dry gulches where flash floods could have deposited the gold; as we walked along "desert pavement," we hoped to find "black gold" (gold, covered with "desert varnish") that would lead us to Peg Leg Pete's Lost Mine.

When we weren't hunting gold, I was hunting books on the subject. I hounded rock shops and bookstores and carried home stacks of library books about gold mining, learned about earth faults, "lode ore," "placers," and the properties of gold from books on mineralogy and geology. Soon mining terms of "glory hole," "dust," "pocket" and "tailings" became a part of our vocabulary.

When "sluice box" began popping up in coversations with fellow miners, we found out what it was . . . had to have one. . . and my research began anew.

I traced my steps back to books, but couldn't find a pattern suitable for us. The Fortyniners had built huge ones, twenty to thirty feet long with "riffles" made of angle-iron, flat pebbles, or poles cut and peeled from nearby trees. They were set in place and used indefinitely. Ours would have to be lightweight and portable, in a size we could carry up and down steep river banks.

We borrowed ideas from the early miners and from folks in camp and tried one after another. After combing, eliminating, redesigning, discarding one idea after another, we finally "got the gremlins out" and built our "Mini-Portable" . . . and it works just great!

(You'll find the plans later in this book.)

As the GOLD BUG bit deeper each year, other aspects of our lives changed. We wanted to travel and be out-of-doors more, so we gave up our car-motel arrangement and bought a station wagon-tent setup and camped as close to the river as possible. But then, each trip to the river showed us we were using up too much time packing and unpacking, so we sold it all and moved into a camper-pickup. Besides the convenience of having the camper packed and "ready-to-go," we then had a complete closet to set aside "FOR GOLD PANNING ONLY."

* * * * *

Each spring, we can't wait to head north. We know that as soon as the snow melts, ghost towns along the Mother Lode

and along other gold-bearing rivers in the West, will come alive with weekend miners and the sound of "GOLD!" will echo from one to another. We look forward to seeing old friends and to sitting around our evening campfires to listen to and exchange tall tales.

Some years we have postponed our trip until fall, when the rivers are at their lowest level (and we can easily get across to the other side); when the scenery is even more beautiful (leaves are turning to reds and golds); and when blackberries are ripe enough to lure us from our quest for gold.

After Labor Day things slow down. Sharp winds skip down the tall mountains to warn us that snow is coming; leaves start to fall; and a stillness settles in as one camper-miner after another waves goodbye and calls out, "See ya' next year!"

THE CHARISMA OF GOLD

GOLD is a remarkable mineral: it can be dented or scratched with a knife blade, or flattened when pounded with a hammer.

It has a low melting point and is malleable and more ductile than any other metal.

It can be hammered into sheets as thin as 1/250-thousandths of an inch, and has been drawn without breaking into a 35-mile wire.

It is heavy: 19.3 in specific gravity.

It is soft: 2.5 on the hardness scale.

It occurs almost everywhere in the world!

Because of its beauty and resistance to chemicals, it has long been prized for use in jewelry, coins, and dentistry.

Because it is eternal and not affected by time, air, water, or other corrosives, the surface of earth satellites is coated with gold to protect them from the heat and corrosion of outer space.

Various industries in our country alone use 8,000,000 ounces of gold every year!

Traces of this abundant mineral have been found in deposits of copper, lead, zinc, coal, and clay.

Even in the sea, scientists have found there are six parts gold for every trillion parts of salt water though no one has found an economical way of retrieving it.

Throughout ALL of history, gold has been considered a "precious" mineral and is probably the first metal to be mined . . . because it can be obtained in its native form with primitive tools.

The Bible refers to "gold" at least 25 times!

Archeologists have uncovered hoards of gold in the ancient ruins of Greece, Egypt, Mexico and Peru; and Egyptian monuments show that gold has been panned, melted, and pounded to fit mankind's wants and needs as far back as 2900 B.C.

On our continent, placer gold occurs in streams and rivers from the Yukon, through the States and down through Mexico.

It has been discovered in almost every state in the Union, from the East Coast to the West Coast . . . so . . . BEWARE! You can catch GOLD FEVER in the desert, in the mountains, or at an ocean beach!

It is lovely and still obtainable.

There is no other mineral so bright, so lustrous, or so astonishingly heavy.

No other metal has captured the hearts of man as has GOLD!

THE DISTINCTIVE BEHAVIOR OF GOLD

In my quest for gold, I discovered it could be found in some places along the river, but not in others, so I searched for the reasons WHY.

The more I read, watched, and listened, the more I realized that the folks who found the most gold . . . knew the most about its nature and character. I wanted that knowledge, too.

My brush with GOLD FEVER had left me with a strong desire to find out the source of "placer gold," but I soon realized I, first, had to back up and study how "lode ore" was formed.

As I continued to read and ask questions, I picked out what I needed to know and tucked important facts away in my mind.

I learned that rocks are NOT solid as the old saying goes: "Solid as a rock," but that they have cracks, pores and cleavages (some so small to be seen only through a microscope), others so huge and deep they penetrate the crust of our earth (called fissures or faults).

When I unearthed that bit of information, I was able to understand what the geologists described: that this ancient metal poured upward from the earth's interior (into fissures in bedrock) and formed in its present state somewhere between two and ten million years ago. That native gold (combined with a few other minerals) is found in "veins" that formed in these fissures when the molecules of molton rock separated and solidified into (mostly) milk-white "bull quartz."

10

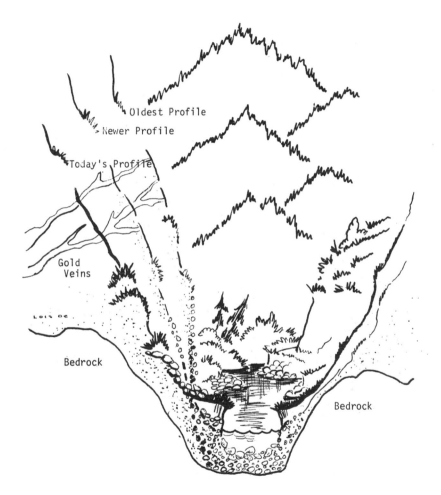

How gold deposits erode, year after year.

This is the "lode ore" that must be mined, milled, and separated. A "placer" is formed when Nature does the "mining," "milling," and "separating" for us.

Whenever ore veins become exposed (by natural erosion of surrounding soil) the bull quartz, in this case, is left unprotected and subject to the effects of the weather.

Alternate changes from summer-heat to winter-cold cause the surface of the quartz to expand and contract and finally to fracture. Along come the early winter rains which penetrate the fractures and is caught there.

As the weather turns colder, the water freezes, expands and enlarges the fractures. This process continues as periodical thaws and freezes act upon the rain-water until the quartz-rock is finally broken free from the matrix rock.

Heavy spring floods start their rampages and rush down mountainsides carrying their collection of rocks to the river. More fragmentation happens along the way as the gold-bearing quartz rolls, bounces, tumbles, and hits other rocks. This releases more of the invulnerable gold.

When these great torrents of water meet the river, the river-water becomes clogged with the burden and the current greedily takes over. Where the turbulence is greatest, the gold particles (being soft and malleable) are pounded into "chunks," "pellets," and "nuggets" by being caused to skip, hit, bump and bang against other rocks.

While the water is running fast, gold and heavy black sand will continue to move along with it; but wherever the current slows down and the pressure eases (at bends in the river), or when the gold meets obstacles (such as huge boulders) it will "drop out" to rest and concentrate. Being a heavy, lazy mineral, gold will follow the easiest and quickest route downstream and stop wherever convenient.

After it drops out and settles to the bottom of the river, the continuing undulation of the water overhead causes the gold to sift down through the "overburden" until it comes to rest on bedrock. Even then it will continue to slide along smooth bedrock until it becomes trapped in a pocket or crevice.

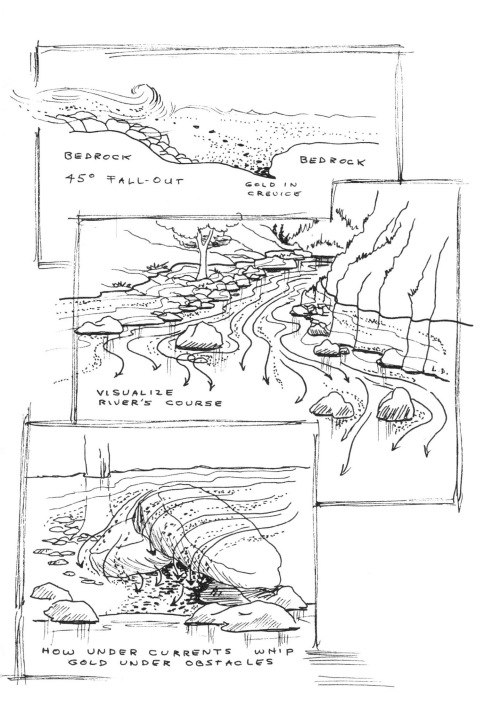

BEDROCK

45° FALL-OUT

BEDROCK

GOLD IN
CREVICE

VISUALIZE
RIVER'S COURSE

HOW UNDER CURRENTS WHIP
GOLD UNDER OBSTACLES

READING & WORKING
THE RIVER

Depending upon the physical nature of a particular river, you may find gold in many types of places; in some other river, in only a few. We've found gold, for instance, along a sandy bank of a river in Colorado that meandered through a meadow (so we could dig ONLY on the sand bars); then again, we've found gold in California along a river where tall, rugged mountains rose from steep-sided banks (could clean out crevices, cut moss, AND dig on sand bars).

So . . . study YOUR river carefully before you start to work.

Take time to visualize the flood-stage and to imagine the course the river would have taken then. Build a picture in your mind of the imaginary high-water line and try to "see" where the gold would have settled out.

Remember that after swift water, gold will "fall out" at about a 45 degree angle.

At sharp bends in the river, it will drop on the inside curve where the flow is slower and the pressure of the current eases.

Look on either side of a huge tree root that may have reached the water during high-flood stages; and along sand bars where the pebbles are the largest.

Old-timers know that the largest pebbles are the heaviest, that they drop out sooner than others; that gold (being heaviest of all) will then fall-out with the heaviest rocks. Coarse gold sinks fastest and first; finer gold is carried farther.

Watch for a sand-laden "cove" which is downstream from

The dotted lines show the old course of the river
before the sand bar built up and diverted the stream.
Anywhere on the sand bar, gold can be found!

some high, protruding bedrock or huge boulders. Both the bedrock and the boulders could have acted as obstacles and caused the current to slow down enough to allow the gold to drop onto the sand.

Occasionally, clues will tell where the original course of a river flowed. Wherever the current slowed down, its burden of sand would settle quickly. After a bar is started, each year more water would be slowed down at that point until gradually the sand bar would build up enough to force the river to take the easier path and move around it.

When this has occurred, gold can be found by digging in the old river-course at the upstream end of the bar.

As you walk along sand bars, watch for arcs of black sand (there will be gold near it), or roll over large boulders (dig where they'd been), or if the boulders are too big . . . GREAT! . . . dig under the downstream side of them. When they were under water, rushing water would have swirled around the sides to form a small eddy which would have slowed the action of the current and, therefore, dropped gold.

Watch for depressions containing the debris of old cans, rusty iron, nails, and/or lead weights (the kind fishermen use). Any one, or all of these can be a clue to the action of an earlier current and is an indication of a natural receptacle for the heaviest of all: GOLD!

Take time to test out an area before settling down to a day's work. A fast check can be made by starting with grass that during flood time had been under water. Pull a clump from the ground and carefully catch all the soil that clings and place all in a pail of water. Work the sand and gravel loose by rubbing, squeezing, pulling apart and dowsing up and down until the grass blades are free.

Pan this "sample" out and you will be surprised at the amount . . . and the size . . . of the pebbles and gold which had become embedded around and within the roots.

If there is "color", there's more gold around and you have a place to begin.

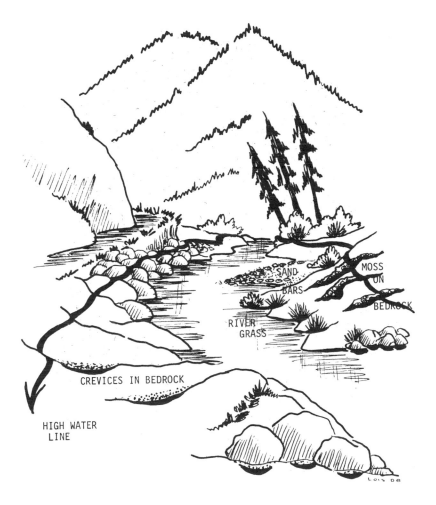

MOSS
ON
BEDROCK

SAND
BARS

RIVER
GRASS

CREVICES IN BEDROCK

HIGH WATER
LINE

Darker areas show where gold may be **found.**

As you begin to find "dust" or "flour" gold and hope to find larger nuggets, try moving upstream. Remember? Coarse gold drops out . . . first; Finer gold is carrier farther.

But if you "go too far" and start to encounter empty pans, go back to where you began and try another tactic. Try to trace the pattern of how the gold got there.

Check out your particles of gold, using your magnifying glass: if they have sharp edges, they haven't traveled very far; if they are rounded and smooth, they have. Ask yourself, "Was this gold dropped here from the current?" or "Did it wash down from the high bank?"

If you decide they have come via the river, then by all means, move upstream where the heavier gold would have settled out first.

If you decide the gold hadn't traveled far, it's possible you are just below the place where the spring floods had washed it down to the river . . . and it may be possible for you to follow the trail of "float" uphill, and who knows, you may trace the gold to its source!

You will sometimes find clay sticking to the material you've dug. It's best then to dump your "diggin's" into a pail of water. Add a drop of liquid detergent (to act as a "wetting agent") then scrub the rocks against each other until the clay is dissolved.

Pour off the muddy water, very carefully, as the gold that still might have the clay-coating will slip away and be lost back to the river.

Keep the properties of gold uppermost in your mind as you study the river and while working:

> it is more than 3 times heavier than the black sand it is found with,
> it "falls out" after swift water has slowed down,
> it works its way down through the overburden, and
> it is always close to bedrock.

Practice patience so as not to overlook any important clue the river gives to you!

It's there for the digging.

It can be yours without filing a claim, without a mining permit (except if you use a dredge) and you can sell it, keep it, or give it away.

Oh, it's a fascinating game of TRYING TO THINK AND OUT-WIT THE RIVER FOR ITS GOLD!

MOSSING

After digging into sand and river banks, you'll want to try your hand at "mossing."

Look around and above you and you'll notice thick, dark green moss growing on bedrock up to the highwater line.

Patches of this moss behave much like little waving fingers to catch fine gold as it tries to rush past when the river is running wild and high.

Imagine what would happen if pieces of shag rug carpeting were in its place. The separate threads would sway under water, like kelp, and the gold would be stopped by the reverse action and fall into and between the strands of carpeting.

The best way to handle it is to cut a slice of moss from the bedrock, using a sharp knife, then placing it carefully in a pail of water. (Don't try to wash too much at a time.) Add a drop of detergent so the fine gold won't float away, then rub the moss between your hands until it falls apart.

Scoop off the floating moss and save.

Slush the sand into your gold pan and work it out, then go back to where you cut the moss and using a whisk broom, sweep the spot thoroughly. Save every bit to pan out. Be sure to clean out small cracks. The gold, as always, would have worked its way to the very bottom!

When you have found gold in your pan, be sure to save the moss it came from. Back in camp, dry it out and burn it in an old frying pan or pot over your campfire. Pan out the residue. Sometimes you'll find more gold in the ashes than you had gleaned from the moss!

CREVICING

Every kind of bedrock has natural cracks, crevices and pot-holes, but the best kind to "snipe" is a crack or break that runs parallel to the river because the action of swirling water would not have eddied the gold out.

Deep crevices cutting across the river will act to trap gold the same as the "riffles" do in your sluice-box.

When you find that "ideal spot," use a pry-bar to open small cracks, then scoop up EVERY BIT of sand until the crevice is clean. Scrape the side-walls with your knife, as gold will sometimes cling to the spot it happened to be when the water dried up.

For larger crevices, after you've cleaned out the sand as far as you can, pour water (roughly) into the bottom to rile up the remaining sand and gold. Then use a "sniffer" and suck up all the water and gold before it has time to complete-ly resettle.

There will be cracks, depressions, and natural riffles in bedrock filled with sand, enough to entice you to go from one to another . . . but if you do, you'll be making it easier for the next fellow with GOLD FEVER; you will have clean-ed off the overburden for him!

Sometimes you will find a crevice packed with a layer of fine gold (or even a handsome nugget) . . . for a "glory hole" can hold more gold than a shoveful of sand!

PEACE AND QUIET AT THE RIVER !

WHAT YOU NEED TO BEGIN

If by now you have a touch of GOLD FEVER and want to open a new door to a wholesome outdoor hobby, you needn't start out with a big investment.

All you need to wear is comfortable, old clothing:

A pair of shorts or cut-off jeans.

Old tennis shoes (that have seen better days) to wear in the water.

Old sox (to keep out sand).

Boots (if you can't stand the cold water).

T-shirts or loose-fitting shirts for men and boys.

A loose-fitting, cool blouse for women and girls.

You can start out by buying just a gold pan. Add a shovel and a few household items and you're ready to hunt the earth's treasure.

It's best to start out on a small scale, learn the secret places where the river has hidden the gold; then as your enthusiasm gathers, dream-up and construct things to enhance your enjoyment . . . add a sluice-box, or even "go all out" and purchase a dredge and wetsuit.

Gold panning is a hobby the whole family can enjoy!

EQUIPMENT FOR PANNING

12" gold pans, one for each person. "Blue" by heating over a fire to remove protective coating.

And, of course, a Shovel!

Two pails are better than one! Buy one smaller than the other, so they will fit together for easier carrying.

Covered plastic bowl.

Three-lens magnifying glass.

Vials to store gold.

Spoons, filed to a point.

Garden trowel

Whisk broom

Sharp knife

Rock-pick

Chisel

Bent Screwdriver

Pry-bar

BUILD YOUR OWN
SNIFFER

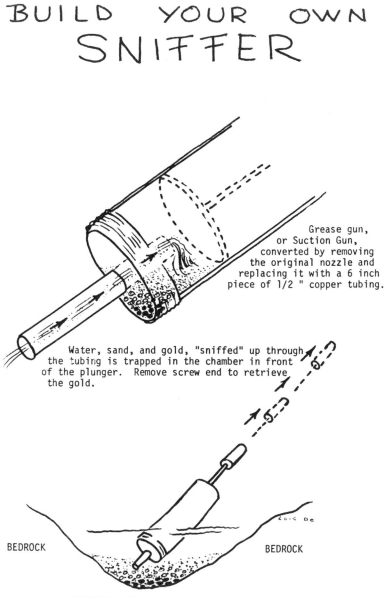

Grease gun,
or Suction Gun,
converted by removing
the original nozzle and
replacing it with a 6 inch
piece of 1/2 " copper tubing.

Water, sand, and gold, "sniffed" up through
the tubing is trapped in the chamber in front
of the plunger. Remove screw end to retrieve
the gold.

BEDROCK   BEDROCK

CREVICE, WITH GOLD AT THE BOTTOM

EASY-TO-MAKE
CARRY ALL BAG

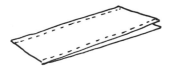

This bag can be made of canvas or cloth-backed plastic.

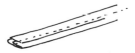

MATERIALS NEEDED:

1 piece 40"x15½" for bag
1 piece 30"x5" for strap

DIRECTIONS

Fold material over with right sides together.

Machine stitch down both sides.

Turn right side out.

Fold down top edge 1½" and top stitch.

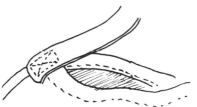

Fold strap piece lengthwise so edges over-lap. Stitch through the 3 layers.

Attach strap to one side, check for size. If the top of the bag, when hanging from your shoulder, comes to about your waist, it will be comfortable to wear.

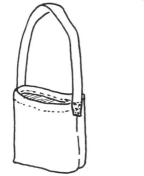

Adjust to your measurement, and sew other end of strap to bag.

Will hold two 12" gold pans and small equipment.

OUR MINING EQUIPMENT

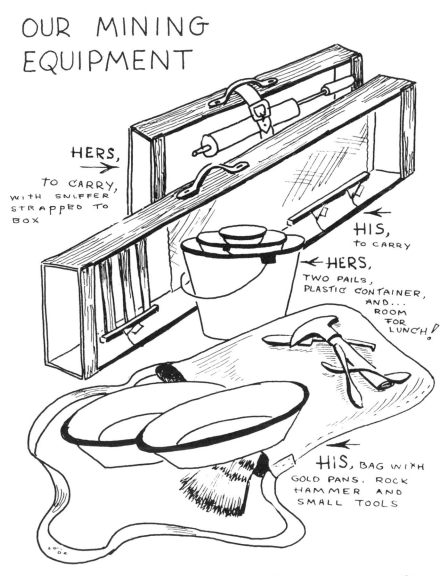

HERS, →
TO CARRY,
WITH SNIFFER
STRAPPED TO
BOX

← HIS,
TO CARRY

← HERS,
TWO PAILS,
PLASTIC CONTAINER,
AND...
ROOM
FOR
LUNCH!

← HIS, BAG WITH
GOLD PANS, ROCK
HAMMER AND
SMALL TOOLS

EASILY PACKED FOR THE WALK TO
THE RIVER

PANNING FOR BEGINNERS

The material you've dug from sand bars and crevices or washed from grass and moss, must be "panned out" to find your gold. Because you'll want to "cover as much ground as possible", it is important to learn to be fast and efficient.

An experienced panner can work from forty to fifty pans a day; so the more you practice, the sooner you'll learn the technique.

No matter what method is used to extract gold, digging a panful at a time, sluicing, dredging, or dry washing in the desert . . . the end process has to be done by panning.

As you run across other gold-miners, watch and practice their methods. Each one will have a slightly different approach and version of how-to-do-it, so try them all. Then develop the method easiest for you.

That is what I have done. The following pages show you my method. It has been successful for me. I hope it will be for you, too!

Once you master a technique, you'll never forget how to use it.

You can practice at home, using an old wash tub, garden cart, or wheelbarrow to hold water. Dig some sand and gravel from your backyard, "salt" it with lead weights or sinkers from your tackle box and begin.

Go through the motions until you can retrieve the sinkers.

When you can do this . . . YOU'RE READY FOR THE RIVER AND ITS GOLD!

Fill your pan about 3/4 full, then submerge it just under the river, keeping it level with the top of the water.

Stir with your fingers to break apart the clay and root particles until everything is separated and moving freely about.

If you've done this thoroughly, your gold will already be at the bottom of your pan.

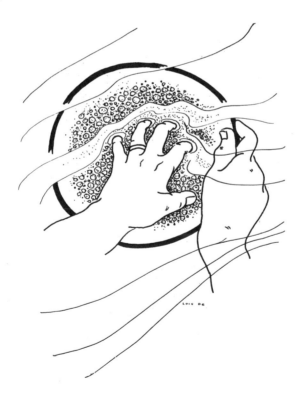

Hold the pan with one hand on each side, dividing the weight equally (still level and under water).

Shake it with quick clockwise and counter clockwise motions to be sure every bit of your material is loose . . . all the way to the bottom.

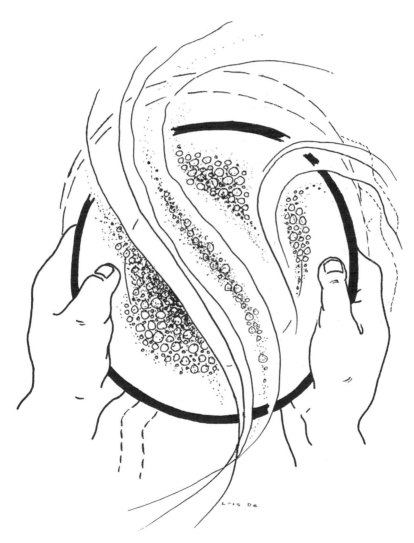

Using a rotating motion in somewhat of an oval pattern, start the material and water to move about.

This motion washes the lighter material off all around the lip of your pan and will give you a feeling the rocks are semi-floating as they move over the edge.

Continue this pattern 3 or 4 times until about half of the overburden is washed away.

Then hold your pan, filled with water, just out of the river and give it sharp clock-wise, then counter-clock-wise motions to resettle the heavy black sand and gold.

Repeat the oval pattern to wash off more overburden, and whenever you think the gold may be working itself to the edge of your pan . . . resettle it.

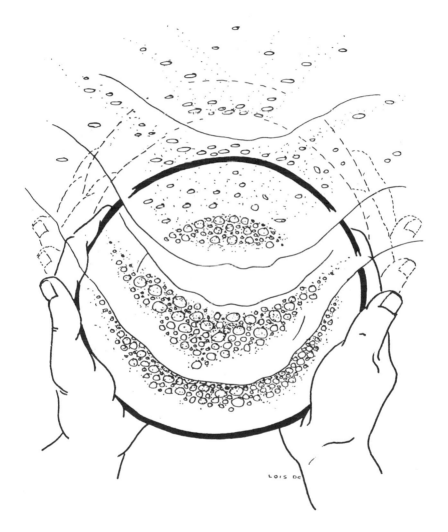

LOIS DC

As the residue of pebbles and blonde sand washes away, the material left in your pan will be less and less, so gradually raise the edge of the pan that is nearest to you. This action will allow the semi-flotation to continue, but now it will be off only the front lip of your pan.

Stop occasionally to pick out pebbles that stubbornly hold on.

When enough of the undesired material has washed away and the weight is less, hold your pan with only one hand and continue the oval rotating pattern, but now, add a jerking forward motion as if thrusting the sand away from you.

Your gold, being heaviest of all, will be underneath the overburden and black sand and will be concentrated at the crease of your pan.

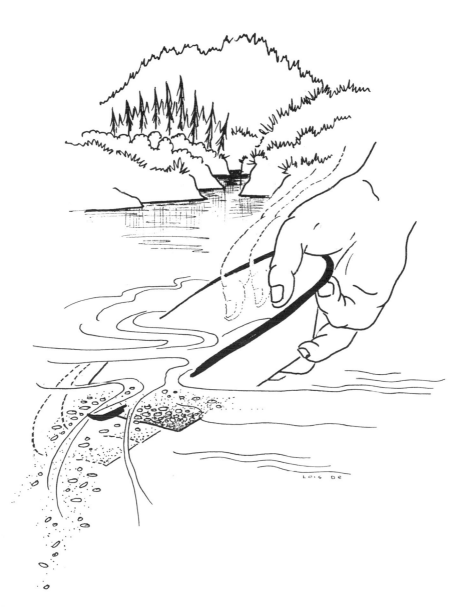

Continue to wash the lighter material over the lip of your pan, shaking and dipping until only black sand is left.

Do not wash out the black sand!

Whenever you see the black sand working its way to the edge, resettle it.

If you have followed this action, so far, it would be almost impossible to lose your gold by accident.

Most of the black sand should be on the flange of your pan, with the heaviest deposit near the crease, and will look somewhat like a fat crescent.

Now, pick up just a little water 1/2 to 3/4 cup, depending upon how much black sand and gold you have left.

Use only enough water to cover your material.

Hold your pan level and tap it hard with the palm of your hand so all the material left is spread on the flat bottom.

This action will re-settle your gold and black sand, and the lighter colored-lighter weight sand will rise to the top again.

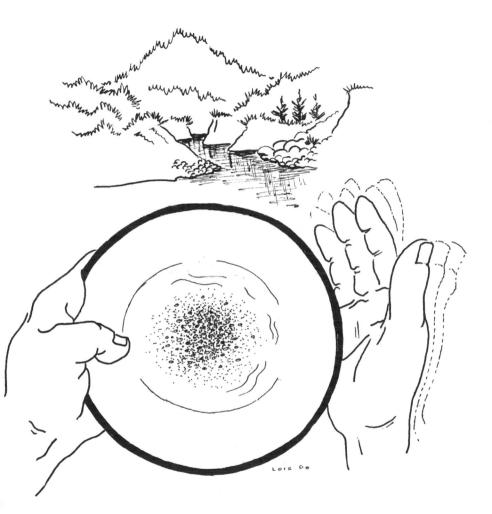

Tilt your pan and start the water to move slowly in a clockwise direction; you will see it start to carry the blonde sands to the left side of your pan.

The black sand and gold (being the heaviest) will trail behind and cling to the right hand side.

Let the water carry the overburden until the black sand starts to follow, then slowly tip your pan until only the waste material is under the water.

Your pile of black sand and gold will cling and stay on the other side.

Many beginners have trouble mastering this phase, but the secret lies in having just enough water in proportion to the sand. The water should have enough weight to move the top blonde sand away; but not so much weight as to push the whole pile around.

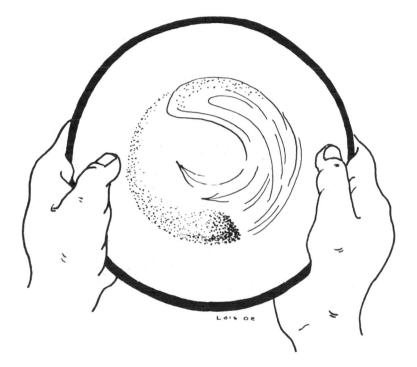

Dip your pan into the river and let ONLY the waste material wash away.

Dip clean water then hit the side of your pan again to settle everything to the bottom again.

Repeat this step until all you have left is black sand.

If you have gold, you will have been catching glimpses of it. Whenever you see flecks appear, resettle the pile and "hide" your gold before washing away the residue.

You could continue this action until there is nothing left in your pan but your gold; but at this point, I prefer to dump the black sand and gold into my plastic bowl and get back to digging more of this alluring gold.

I can always finish the separation in camp, or even later, at home.

Lois De

Wing
Dam

Rock to anchor
sluice

Bed of
rocks

Tailings

Lois oe

MINI-PORTABLE
SLUICE BOX

After a season or two of panning, you'll want to "graduate" to using a sluice-box because you can then process 5 to 10 times the amount of ground.

The more ground you process, the more gold to carry home!

Depending upon the location and the amount of sand and gravel available (and where there is sufficient current), sluicing is the most economical method of working placer ground.

To understand the proficiency of a sluice, think of it as an imitation of the river.

Just as the river traps black sand and gold with its natural riffles, the sluice catchs and holds the gold, but in a concentrated way. It reproduces in a small area what the river does on a large scale.

Our Mini-Portable Sluice channels the river water (through an open-ended trough) over metal lath (which acts like cracks or crevices), over carpeting or toweling (which traps the same as moss), and over wooden riffles (which copies natural riffles in bedrock).

You'll find it a pleasure to use. It is practical and simple to construct and operate.

Once you get your sluice set up, then you are free to dig pailsful of ground at a time; just dump the sand in your sluice and let the river do the separation for you while you go back to your diggin's for more!

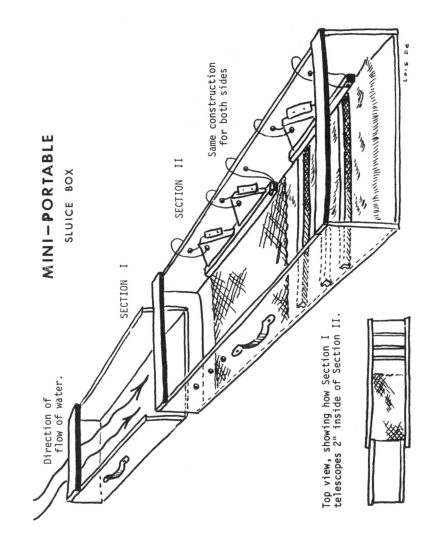

MINI–PORTABLE
SLUICE BOX

Direction of
flow of water.

SECTION I

SECTION II

Same construction
for both sides

Lois De

Top view, showing how Section I
telescopes 2" inside of Section II.

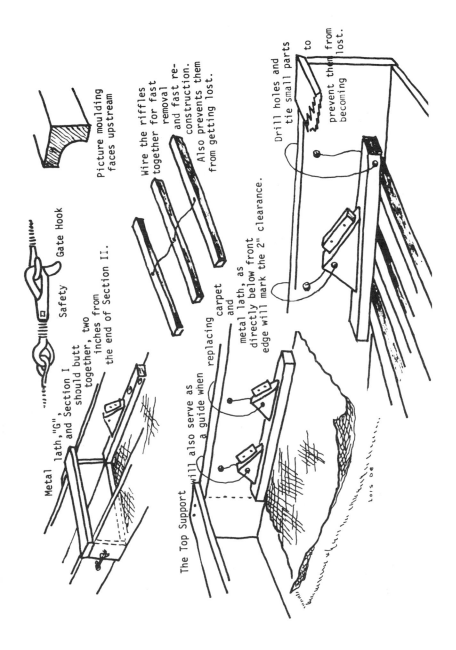

Picture moulding faces upstream

Metal lath, "G", and Section I should butt together, two inches from the end of Section II.

Safety Gate Hook

Wire the riffles together for fast removal and fast re-construction. Also prevents them from getting lost.

The Top Support will also serve as a guide when replacing carpet and metal lath, as directly below front edge will mark the 2" clearance.

Drill holes and tie small parts to prevent them from becoming lost.

Lois De

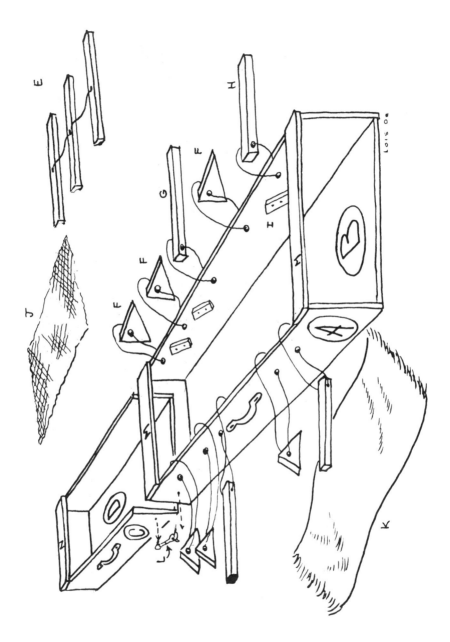

LOIS OB

MATERIAL TO ORDER FOR SLUICE

Amount	Size	Description	Part
2	3/8 x 4"x4'	Two side panels	A
1	3/8 x 12" x 4'	Bottom floor	B
2	3/8 x 3-5/8"x2'	Two side panels	C
1	3/8 x 11-1/8"x2'	Bottom Floor	D
3	1/2"x 1/2"x12"	Picture Moulding	E
6	1" x 1" x 5"	Wedges	F
2	1" x 1" x 28"	One for each side	G
2	1" x 1" x 10"	One for each side	H
6	1" x 1" x 3"	Braces	I
1	12" x 28"	Galvanized metal lath	J
1	12" x 40"	Towel or carpet	K
2		Safety gate hooks	L
2	3/8"x2"x12-3/4"	Top braces Section II	M
1	3/8"x2"x11-7/8"	Top braces Section I	N

USING YOUR SLUICE

Before placing your sluice in the river, look for a fast-moving current close to the bank where you will have easy access to both your sluice and your source of gold-bearing material.

Arrange a "bed" for it by placing rocks in the river in such a way as to support the entire length. Leave a drop-off at the end for the tailings to fall, where they can easily be pushed away when they accumulate.

Regulating the water passing through is very important.

Too fast and too much water will wash EVERYTHING away, including your gold; too little current will allow the riffles to become clogged and the tailings to pile up.

A good rule-of-thumb is to set your sluice at an angle, deep enough so the water will pass through about half way up the sides of the upstream end (about 2 inches deep). There should be enough of a down grade to carry off the overburden out the lower end (about an inch from the top).

Place a large rock on top of each support to keep your sluice in place.

A wing-dam can be made of rocks to divert faster water when needed.

When your sluice is set up to your satisfaction, you're ready for production.

Pour the placer ground you've dug into the back section, stir with your hand until ALL of it is thoroughly wet (even fine gold will float off when dry), break up clumps of clay, then allow the river to carry it over the metal lath and riffles.

You will be able to watch the lighter pebbles and rocks rush forward with the current and see how little sand bars of black sand are left behind.

In the black sand, you'll see flecks and flakes of gold which will gradually wash on and become lodged in the carpeting and riffles. (Take time, sometimes, to watch this happen and you will have a better understanding of how the gold moves in the river.)

You can continue to dump material into your sluice until you think the carpet will hold no more. You may have to clean it every hour or so, or you may let it go a half a day. It will all depend on the amount of black sand you have found.

Check the screen before dumping the next pailful of dirt to see if some flakes are large enough to be removed with a pair of tweezers. If so, stop the flow of water before trying to capture your gold. If you try to pick up the gold when the water is moving past, the tweezers (or your fingers) will cause an eddy and the gold will whiz on past.

When you are ready to "clean up," stop the flow of water and lift your sluice to dry land. Remove the wedges, supports, riffles, metal lath and carpeting.

Wash each piece carefully over a bucket of water. When everything else has been rinsed, it's time to do the carpeting. Lift it carefully so you will save whatever sand is clinging to it. Wash thoroughly in a full bucket. Rinse and squeeze and dowse up and down. If the water becomes muddy, pour it off slowly and fill the pail again with clean water.

Keep washing out the carpeting as that is where most of the gold will be caught.

Before re-assembling your sluice, hold it over a pail and pour water down each side to clean any gold which may have stuck to the wood; put it all back together and back on its "bed" and you are ready to pan out the material in your pail.

CLEANING UP SLUICE

"SLUSHING"

You will find that the fine sand and gold will cling to the bottom of a pail and will not wash out by just rinsing and dumping it. No matter how many times you rinse and dump, there will be some left.

So, set your gold pan in shallow, quiet water and hold your pail so the bottom edge is directly over the center of your pan.

With your free hand scoop water and "slush" it up to the inside bottom of your pail.

Do this two or three times and all the material will slide down into your pan and the pail will be clean.

After panning your material down as far as you care to, use this same "slushing" action to transfer the black sand and gold into your plastic bowl.

SEPARATING GOLD
FROM SAND

Every "Gold Bug" has his favorite method of retrieving his gold, and you will hear of many ways; but I have found that the simplest way is to start by pouring some of the wet gold mixture into a gold pan.

Spread a thin layer, then set the pan on the edge of your campfire to hasten the drying. Do not let it get too hot.

When the sand and gold is dry and loose, dump about a half cup onto a paper plate.

Place a small magnet inside a paper cup (plastic cup or wrap in a piece of paper), then hold the cup above your paper plate and move, slowly, above your black sand.

The magnetic sand will jump to the magnet (protected by the paper) and form long black fingers. Before you lift it away, tap the cup lightly to release any fine gold that may have been trapped with the sand.

Do not use your gold pan in place of the paper plate. The magnet will grab the metal pan and take up too much sand (maybe gold, too) at a time.

Lift the magnet away and lift it out of the cup. The particles will drop off the paper and your magnet will be clean.

Repeat until you have removed all the magnetic sand.

Whatever black sand is left can be moved aside or blown away very carefully.

Now you can really see your gold and can pick it up to store in a small bottle by using a pair of tweezers, a small wet artist's brush, or a wet wooden match stick.

56

Fill your small vial with water before you start, then as you touch the top of the water with your brush carrying the flecks of gold . . . the gold will drop off and settle, quickly, to the bottom.

Gold looks larger through the water . . . a good morale builder!

Some vials are made especially for this purpose and may be purchased at hardware stores in most old mining towns (wherever gold pans are displayed) or at your favorite rock shop before leaving home.

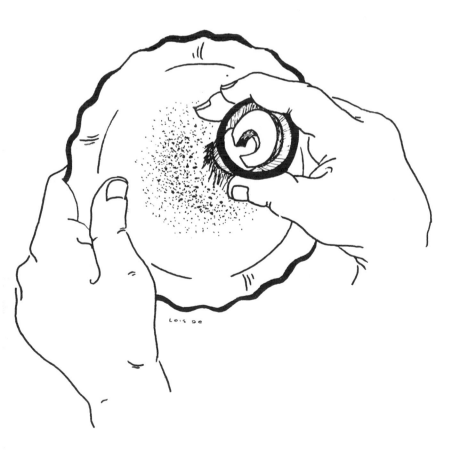

Lois De

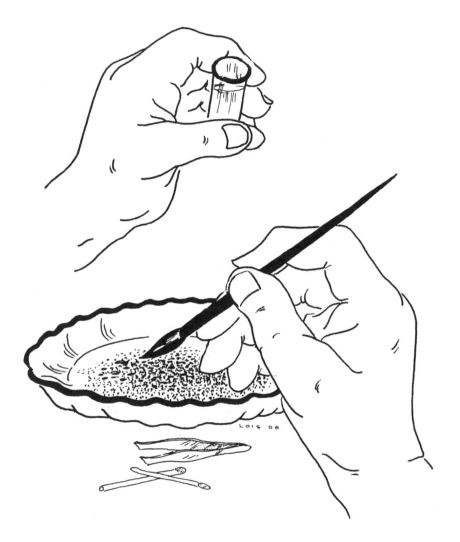

USING MERCURY

If you want to use mercury to recover your fine gold, place a little in a batch of wet (or dry) black sand mixture and let it roll around until all the fine gold is taken up and the mercury looks fat and floppy.

Place this blob into a small piece of chamois (even a handkerchief will do), gather all the edges and twist until the mercury oozes through the pores (or the threads) and leaves a button of silver-colored amalgamated gold.

Using an old "throw away pan," heat the button slowly over an outside fire. The mercury will evaporate and leave the gold free. Let cool a bit, then slowly add vinegar to cover. Bring to a boil for 3-4 minutes, pour off vinegar and wash with soapy water.

CARE MUST BE TAKEN! DO IT OUTSIDE!

DO NOT MELT THE GOLD.

DO NOT USE THE PAN FOR ANYTHING ELSE!

DO NOT BREATH THE FUMES!

MERCURY FUMES ARE DEADLY POISONOUS!

Old-timers used potatoes. They sliced the potato in half, scooped out a hollow center big enough to hold the amalgamated gold, then tied it with wire (could be wrapped in foil) and laid it in a bed of coals to bake an hour or so.

The mercury becomes absorbed into the potato and the gold will be shining and free.

Drop the potato immediately into a pail of cold water. The beads of mercury will ooze out and drop to the bottom to be retrieved and used again.

BE SURE TO DISCARD THE POTATO REMAINS SO NO PERSON OR ANIMAL WILL EAT IT!!! DO NOT BURY IT!!

Fine gold can also be salvaged by mixing water and oil together in a container. Shake well, then let the black sand settle. The oil will rise to the surface bringing the gold with it.

(This should be the ONLY time you use oil near gold.)
This process is called "frothing." Gold re-agents and froth-ers can be purchased from chemical companies very reason-ably priced.

MAKING GOLD JEWELRY

If you want to find ways and means of using your gold, so others may enjoy it too, beautiful gifts and heirloom treasures can be made at home.

Hollow pendants may be purchased at jewelry shops or rock shops for a very reasonable sum then filled with flakes and fine gold. Be sure to work over a piece of white paper. Open the pendant and fill both sides, then quickly put back together. Add the ring and tighten the screw-lock and you have a wonderful gift worth $25 or more.

A nugget can be left "as is," to have a jeweler add a jump ring. It can be hung on a chain for a lovely necklace. If you have two that fairly match, how wonderful to make a pair of earrings from nuggets you've found for yourself!

To make attractive plastic jewelry embedded with gold, start by hammering a few flakes at a time to make them appear larger.

Gather the following equipment:

> Casting resin
> Catalyst (hardener)
> Dyes
> Lacquer thinner (for cleaning up tools)
> Molds
> Paper cups (for mixing)
> Flat sticks (for stirring)
> Toothpicks (for adding color)

Clock (for timing)
Heavy paper to cover work area
Small artist's brush (to pre-coat gold)
Measuring spoons
Sandpaper (to smooth bottoms if necessary)
Tweezers

Directions for using the resin, dyes and molds can be purchased along with your supplies at hobby shops, some hardware-paint stores, and at most rock shops.

Before making your finished products, practice making small batches of resin to become familiar with the product and the action of the catalyst. After you have practiced pre-coating something similar to your gold (old sequins for instance), be sure you have the knack of coating before starting to work with your precious gold.

Choose a small mold as for cabochons. Mix a small batch of resin and catalyst and pour in a clear "base." It should be just deep enough to cover the bottom of the mold, no more than 3/16" thick.

Let it set or "gel".

Place your coated gold on top where it will make an attractive pattern, remembering that you are working upside down. The clear base will be the finished top, so visualize what it will look like when turned over.

Mix another small batch of resin. Pour in the second layer with just enough resin to cover the gold and to "anchor" it in place.

Let it set or "gel."

Your last layer should have color, so add a little dye to the resin by dropping coloring off the end of a toothpick BEFORE the catalyst is added. Do not use too much color or it will delay proper hardening later.

Green, blue and turquoise colors will show off the gold better than other colors.

Remove the piece from the mold by turning it over and tapping it gently. If necessary, rub the flat back across sand-

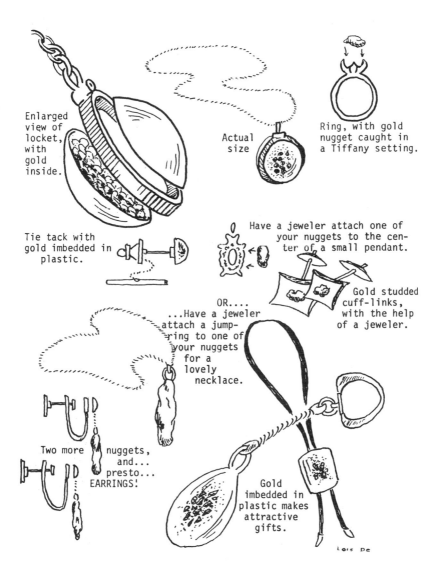

Enlarged view of locket, with gold inside.

Actual size

Ring, with gold nugget caught in a Tiffany setting.

Tie tack with gold imbedded in plastic.

Have a jeweler attach one of your nuggets to the center of a small pendant.

Gold studded cuff-links, with the help of a jeweler.

OR....

...Have a jeweler attach a jump-ring to one of your nuggets for a lovely necklace.

Two more nuggets, and... presto... EARRINGS!

Gold imbedded in plastic makes attractive gifts.

Lois De

paper lying flat on the table. Never rub the top.

It will take from 5 to 7 days for the piece to cure completely, so handle gently and set it aside.

Parts to make key-chains, earrings, pendants, etc. are called "findings" and can be purchased at rock shops.

WHERE TO SELL IT

If you can part with your gold, the yellow pages of phone books in most large cities will list the following:

JEWELERS . . . they like to purchase speimens, but will buy flour gold and nuggets of all sizes.

GOLD PLATERS . . . use gold to melt and reuse in various endeavors.

LICENSED GOLD BUYERS . . . are not interested in specimens and pay at current monetary value.

MUSEUMS . . . are always interested in unusual nuggets, so you may recover a good deal more than the acutal weight-worth.

ROCK SHOPS . . . buy for resale, so some nuggets and gold flakes are worth more depending on their shape and design.

If you want to weigh what you have:

24 Grains	=	1 pennyweight
20 pennyweight	=	1 troy ounce
12 troy ounces	=	1 troy pound

If, over a period of time, your gold acquires a film, as even fine jewelry does, it can be brightened in a mixture of vinegar and water with a bit of baking soda added to hasten the action.

Using an old toothbrush and some toothpaste will do the trick, too!

GOLD PANNERS
MINING TERMS

BAR . . . pertains to rivers, generally a sand bar.

BLACK SAND . . . is usually magnetite, horneblende, hematite and other minerals. It is very heavy and is found with gold.

DIGGIN'S . . . for weekend miners, it refers to wherever they happen to be working along the river on any particular day.

DREDGING . . . is a method of processing a great deal of material from under water. A suction hose acts much like a vacuum cleaner to pick up sand, water and gold from the river. A motor and sluice floats in mid-stream atop a large inner-tube. Operators wear wet-suits, goggles and other scuba gear.

DRY-WASHING . . . is the process used in the desert when no water is near. Gold-bearing sand is shoveled into the "hopper" at the top, then billows are worked, either by hand or by a motor, and the overburden is blown off by air-power, leaving the gold to settle in riffles. The final step of retrieving the gold is panning.

DUST . . . a term used to describe minute particles of gold.

FLOAT . . . the particles of gold that stays on the surface of the ground, usually scattered downhill from its original source.

FOOL'S GOLD . . . is mica, pyrite, or cubes of iron pyrite. All will crack and crush and, under a magnifying glass, will appear rough and grainy. Will not shine in the shade as gold does.

GLORY HOLE . . . is any crevice or low spot where you've found a concentration of nuggets or flakes.

GROWLERS . . . are nuggets which can be heard "growling" in your pan even before you get down to seeing them.

GOLD . . . will shine in shade or sunshine. It can be of different colors. When alloyed by nature with silver, it will appear lighter in color; when alloyed with copper, it will appear darker.

HARDROCK MINING . . . is what vacation-miners call it when having to use a pick, pry-bar, gads or rock hammer to open crevices.

HI-GRADING . . . is the expression used when finding a nugget just lying on bedrock, or loose on the ground where no digging is necessary. In the old days, it meant miners were illegally removing gold from a mine where they were employed.

NUGGET . . . a lump of native gold of no special size as long as it "rattles" when you shake your gold pan.

OVERBURDEN . . . the silt, sand, and gravel that accumulates at the bottom of a river bed above bedrock and gold.

PANNING . . . a method of extracting gold from stream beds.

PLACER . . . where gold is found with concentrations of sand, gravel, silt, and/or boulders.

PLINKERS . . . very small nuggets, but big enough to be picked up with tweezers.

PLUNKERS . . . nuggets that make a "plunk" sound when dropped in your empty pan.

POCKET . . . generally means a low spot or hole or crevice in bedrock that has captured dust and nuggets.

SLUICING . . . a method of extracting gold from a river by using a series of troughs, with riffles or slats attached to the bottom to trap gold.

SNIPER . . . the name given to a person who uses a "sniffer" and snipes gold from under water crevices.

TAILINGS . . . or mine-dumps are mounds of earth left after gold-bearing ore has been removed. Tailings occur at the end of a sluice-box and must be pushed away periodically.

VEIN . . . a term referring to a lengthy occurrence of an ore.

WHERE TO FIND IT

In California, gold can be found in most rivers and their tributaries from Siskiyou County to San Diego County; but we love best the Gold Country (almost 200 miles of it), which extends from the foothills of Yosemite, along Highway 49, to the swift-flowing rivers of the Sierra Nevadas.

BUT . . . you could contract GOLD FEVER along rivers and creeks in Alaska and in our other Western States . . . from Arizona and New Mexico to Oregon, Idaho, and Montana.

So . . . load up your "covered wagon" and head out for adventures galore!

CALIFORNIA

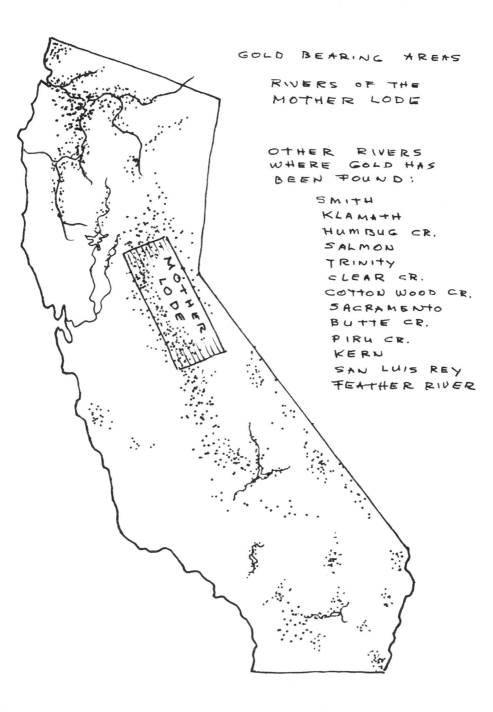

GOLD BEARING AREAS

RIVERS OF THE
MOTHER LODE

OTHER RIVERS
WHERE GOLD HAS
BEEN FOUND:

SMITH
KLAMATH
HUMBUG CR.
SALMON
TRINITY
CLEAR CR.
COTTON WOOD CR.
SACRAMENTO
BUTTE CR.
PIRU CR.
KERN
SAN LUIS REY
FEATHER RIVER

MOTHER LODE

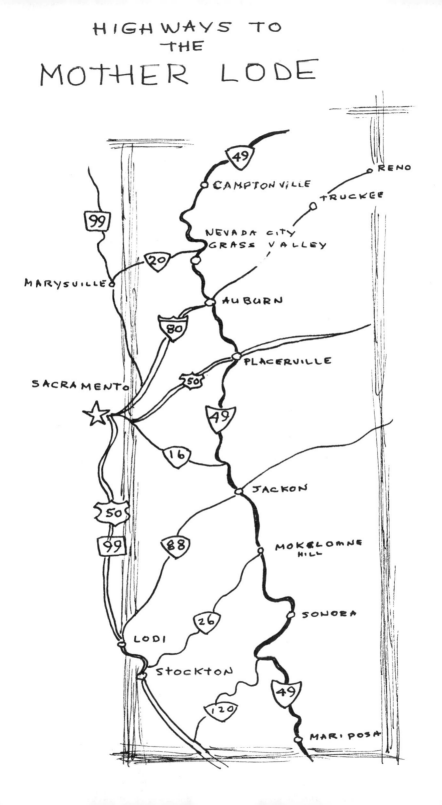

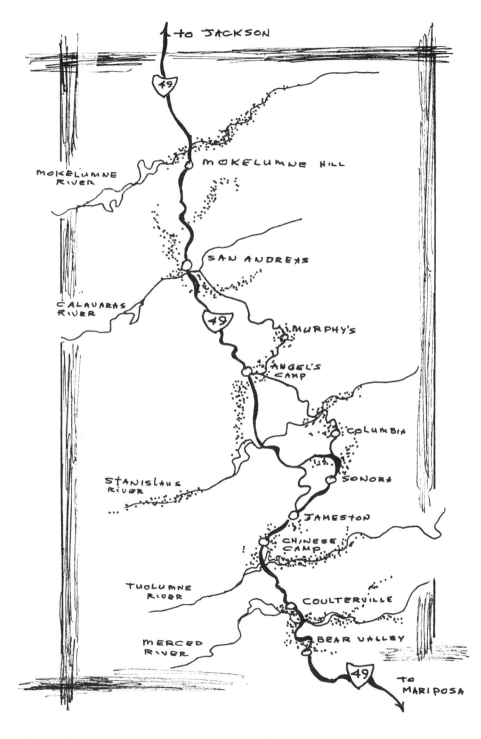

CALAVERAS, TUOLUMNE AND MARIPOSA COUNTIES

to JACKSON

49

MOKELUMNE HILL

MOKELUMNE RIVER

SAN ANDREAS

CALAVARAS RIVER

49

MURPHY'S

ANGEL'S CAMP

COLUMBIA

STANISLAUS RIVER

SONORA

JAMESTON

CHINESE CAMP

TUOLUMNE RIVER

COULTERVILLE

MERCED RIVER

BEAR VALLEY

49

to MARIPOSA

PLACER AND ELDORADO
COUNTIES

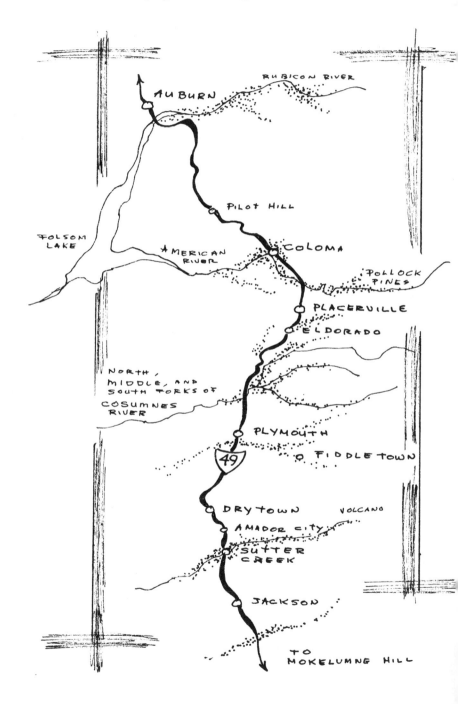

AUBURN

RUBICON RIVER

PILOT HILL

FOLSOM LAKE

COLOMA

AMERICAN RIVER

POLLOCK PINES

PLACERVILLE

ELDORADO

NORTH, MIDDLE, AND SOUTH FORKS OF COSUMNES RIVER

PLYMOUTH

49

O FIDDLETOWN

DRYTOWN

VOLCANO

AMADOR CITY

SUTTER CREEK

JACKSON

TO MOKELUMNE HILL

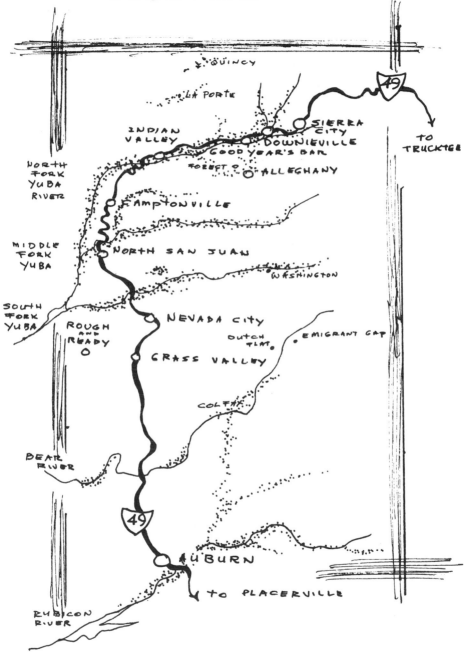

SIERRA AND NEVADA COUNTIES

to QUINCY

La PORTE

49

SIERRA CITY

INDIAN VALLEY

DOWNIEVILLE

GOODYEAR'S BAR

FOREST

ALLEGHANY

to TRUCKEE

NORTH FORK YUBA RIVER

CAMPTONVILLE

MIDDLE FORK YUBA

NORTH SAN JUAN

WASHINGTON

SOUTH FORK YUBA

ROUGH AND READY

NEVADA CITY

DUTCH FLAT

EMIGRANT GAP

GRASS VALLEY

COLFAX

BEAR RIVER

49

AUBURN

to PLACERVILLE

RUBICON RIVER

WASHINGTON —
OREGON

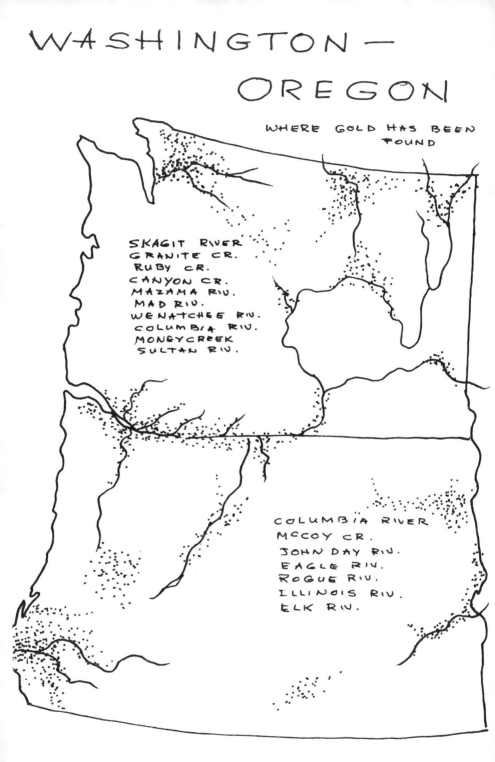

WHERE GOLD HAS BEEN
FOUND

SKAGIT RIVER
GRANITE CR.
RUBY CR.
CANYON CR.
MAZAMA RIV.
MAD RIV.
WENATCHEE RIV.
COLUMBIA RIV.
MONEYCREEK
SULTAN RIV.

COLUMBIA RIVER
McCOY CR.
JOHN DAY RIV.
EAGLE RIV.
ROGUE RIV.
ILLINOIS RIV.
ELK RIV.

IDAHO - NEVADA - UTAH

WHERE GOLD HAS BEEN FOUND

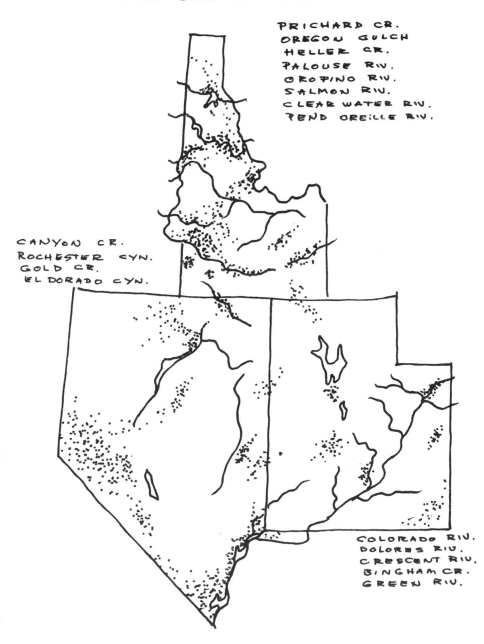

PRICHARD CR.
OREGON GULCH
HELLER CR.
PALOUSE RIV.
OROFINO RIV.
SALMON RIV.
CLEAR WATER RIV.
PEND OREILLE RIV.

CANYON CR.
ROCHESTER CYN.
GOLD CR.
EL DORADO CYN.

COLORADO RIV.
DOLORES RIV.
CRESCENT RIV.
BINGHAM CR.
GREEN RIV.

MONTANA—
WYOMING—
COLORADO

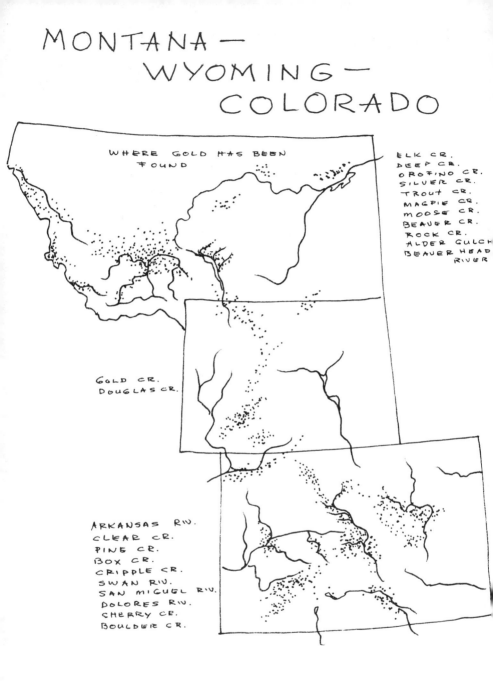

WHERE GOLD HAS BEEN
FOUND

ELK CR.
DEEP CR.
OROFINO CR.
SILVER CR.
TROUT CR.
MAGPIE CR.
MOOSE CR.
BEAVER CR.
ROCK CR.
ALDER GULCH
BEAVER HEAD
RIVER

GOLD CR.
DOUGLAS CR.

ARKANSAS RIV.
CLEAR CR.
PINE CR.
BOX CR.
CRIPPLE CR.
SWAN RIV.
SAN MIGUEL RIV.
DOLORES RIV.
CHERRY CR.
BOULDER CR.

ARIZONA — NEW MEXICO

WHERE GOLD HAS BEEN FOUND

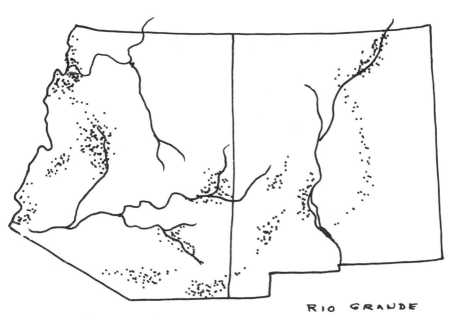

BURRO CR.
BOULDER CR
GRANITE CR.
GILA RIVER
HASSAYAMPA RIV
CAVE CR.

RIO GRANDE

AREAS:
ELIZABETHTOWN-
BALDY
MOGOLLON
LORDSBURG

IF I'VE MISSED A SPOT....
WHO KNOWS...IT JUST MIGHT
BE THAT IT HASN'T BEEN
DISCOVERED YET!

POSTSCRIPT

Looking for and finding gold is not an easy, comfortable, homey type of recreation. It is hard work, but it will get you out-of-doors and reward you with invigorating, adventure-filled outings and escapades. It will stir and awaken the dormant pioneer blood that is inherent in most of us.

You will become so engrossed, you will not be aware of your cold feet (from icy mountain streams), skinned knees (from scrambling over boulders), broken finger nails (from digging in crevices), or weary bones (from shoveling sand and gravel), or even bites from mosquitos.

When you first start out, you will be a little stiff and sore from using forgotten muscles, but as the days accumulate and your store of gold increases, you will forget your scrapes (add a few more bandaids, rub on liniment, and spray more insect repellent). You'll be up with the sun and eager to be on your way to compete with the river.

As you get acquainted with the river, it will seem to take on a personality of its own and will appear to be jealous of her gold; so you'll have to keep your mind sharp, your eyes alert, and your feet ready for action.

You'll soon be climbing over rocks and boulders that had loomed as huge obstacles to you only a few days before, and you'll find the secret hiding places wherever the river has left her clues.

Then, only when the tired old sun has used up its strength and has dropped behind the mountain, will you give up and return to camp to rest and relax.

A rip-roaring campfire, a warming beverage, friends to visit with, and a quiet star-spangled night will bring back renewed energy and you will be as fresh as the new sun when it comes up in the morning.

Surrounded by the eternal mountains, rivers, and skies, your small anxieties that had seemed so pressing will be minimized and placed in their proper perspective; and when it's time to return home, you'll have renewed vigor to face reality.

I can't think of a healthier hobby. You make new friends, take "time out" to fish or swim, or hike up a pine-covered path to an abandoned mine, or peek into a cave where old ore-tracks emerge.

You will have gathered happy memories, which you will treasure more than your gold.

* * * * *

Before you get involved, I want to warn
you. When you've caught
GOLD FEVER:

You will be haunted by every river and stream, thinking "there may be gold there."

You will look for GOLD in the mountains and in the deserts.

You will plan more outings which will be more productive and enjoyable.

You will talk about it with everyone who will listen.

You will search out other GOLD FEVER victims.

You will confide in each other and exchange ideas.

You will enjoy listening for hours around campfires, of tall tales of hidden treasures.

You will be in awe of the size nuggets found by the skin-divers and dredgers.

You will relish with delight to hold these heavy nuggets in your hand.

You will always be watching for black sand, because you know that GOLD will be near.

You will have a constant hope that the next pan of sediment will be "pay dirt."

You will find "just one more crevice" to dig out.

You will believe the old saying, "The river is laughing at you." It's saying, "Ha, Ha, I didn't hide it under THAT rock, I hid it under THIS one." And . . . you'll keep hunting until you find it.

You will want to know more and more about geology and how the GOLD was formed.

You will read about the FORTYNINERS and how THEY found theirs in the old days.

You will pore over books and maps in libraries.

You will prowl around old abandoned Ghost Towns and become saturated with the LURE of GOLD.

You will forget all other worries and problems in your search.

You will see new beauty in the mountains, the laughing waters, the flittering butterflies, the chattering birds, and be delighted by the colorful wild flowers.

and . . .

You will never mistake pyrite for GOLD ever again, once you've seen the real thing in your pan!